FRACTALS (TO COLOR) BY 5LEAF

25 Mathematical Designs for Focus and Concentration

©2016 S.A. Schlager. All Rights Reserved.
FractalsBy5leaf@gmail.com
www.Fractalsby5leaf.com

Feel free to use this page as a "test page" for your pens, markers, pencils, etc. so you don't have to spoil any of your fractals!

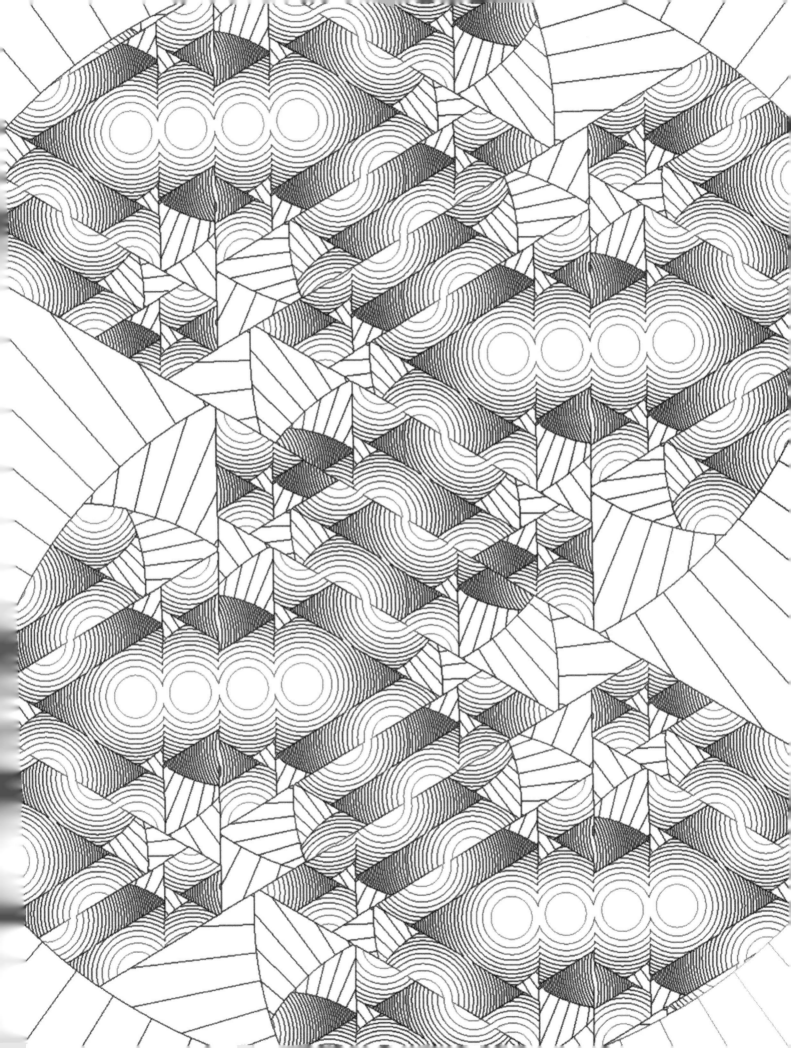

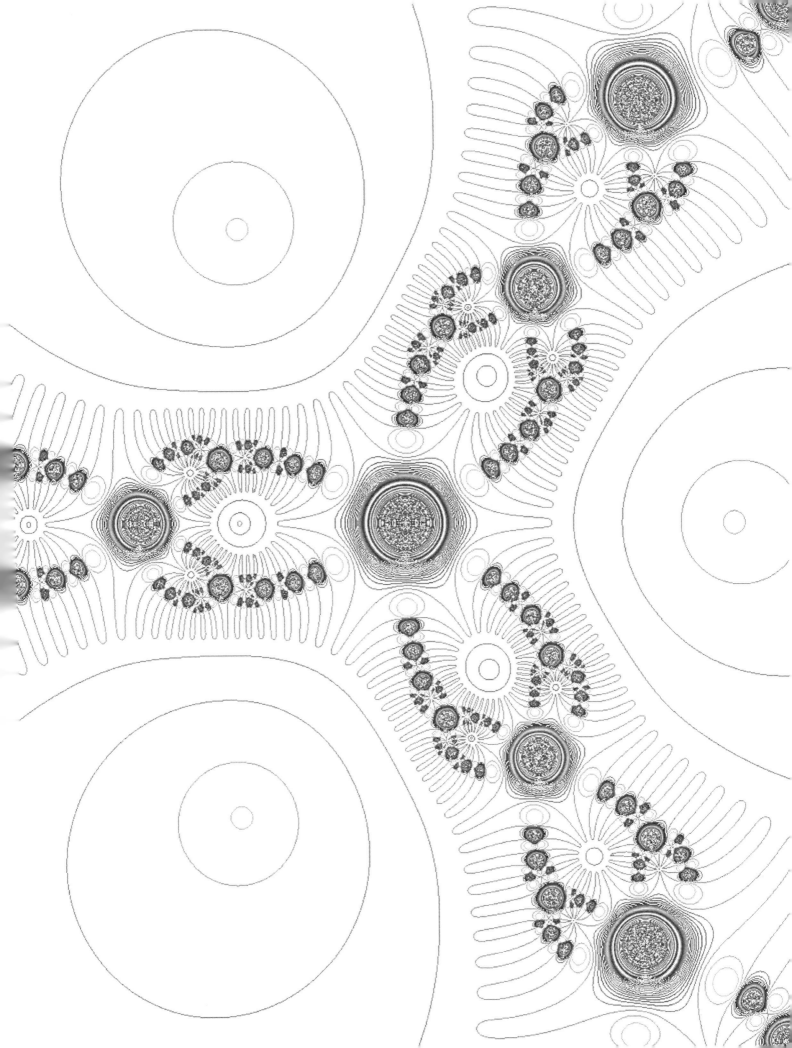

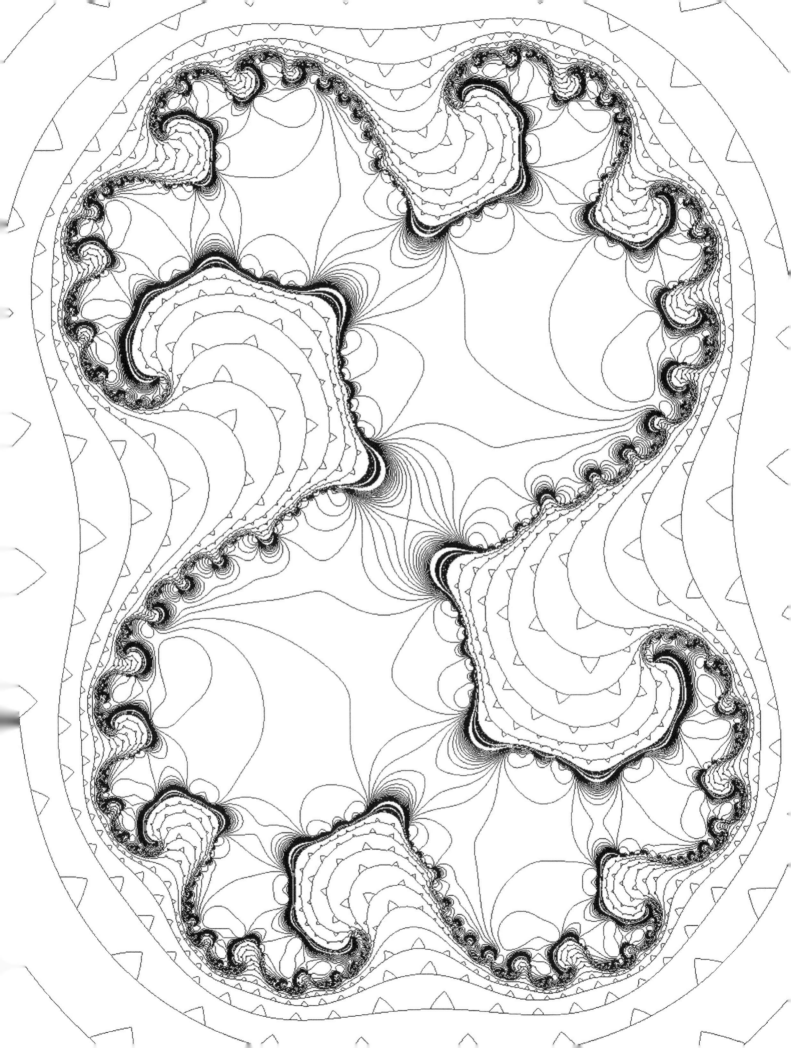

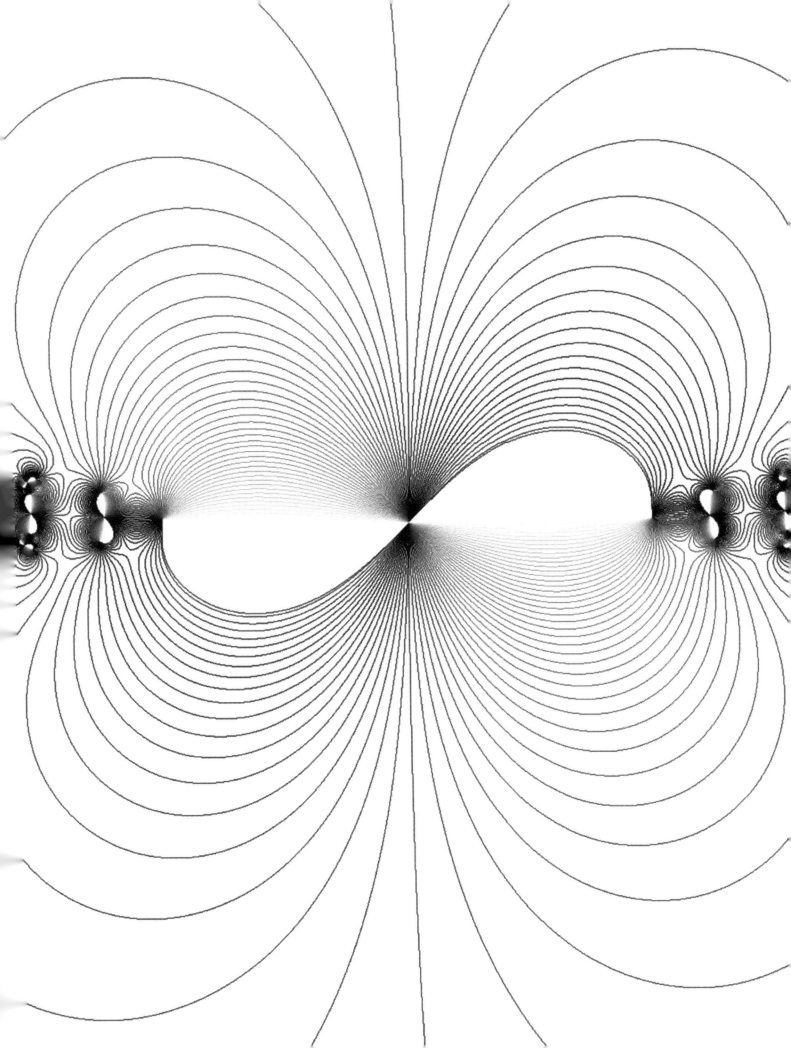

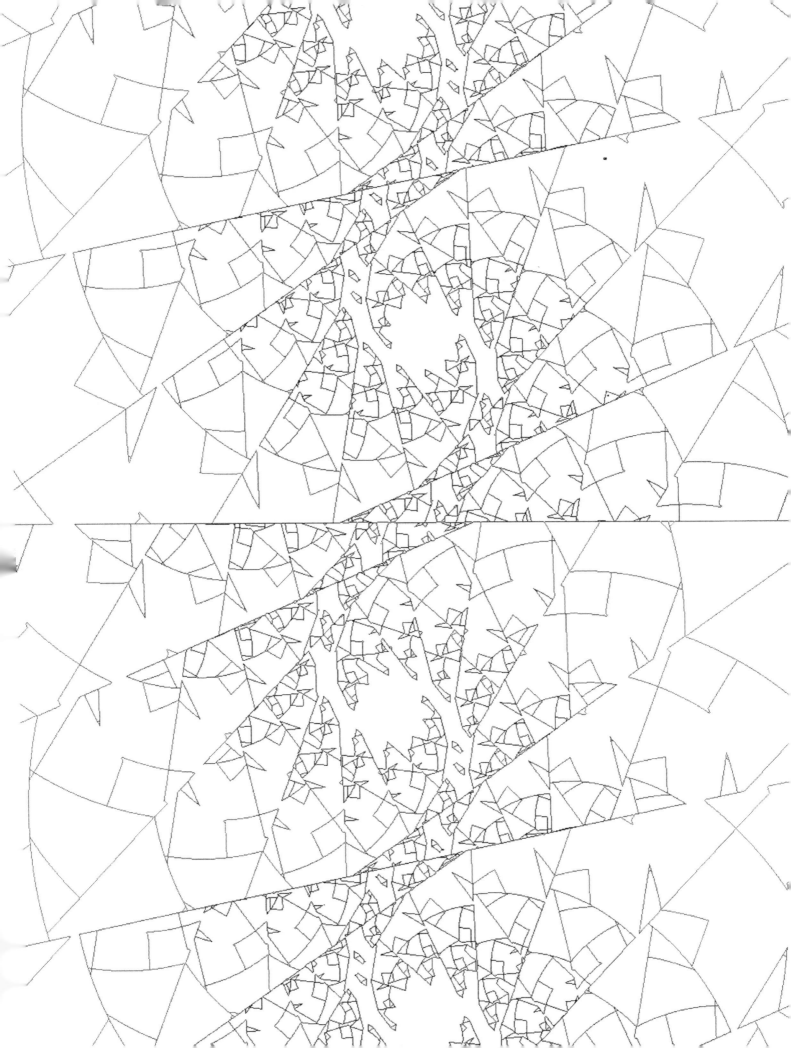

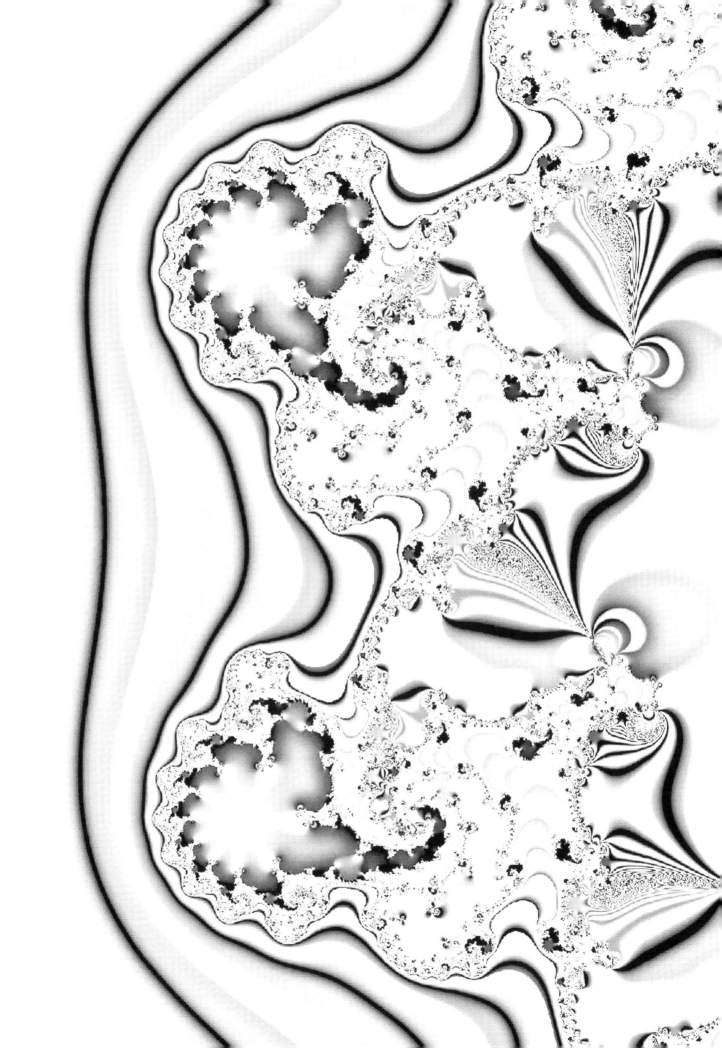

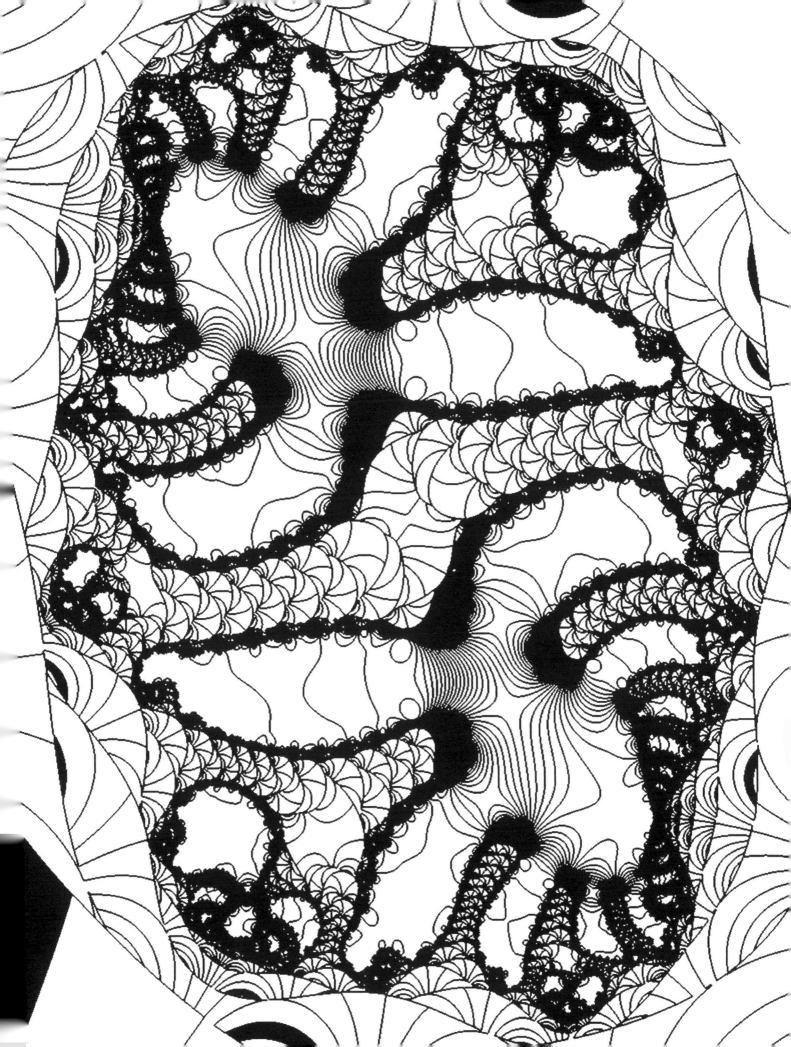

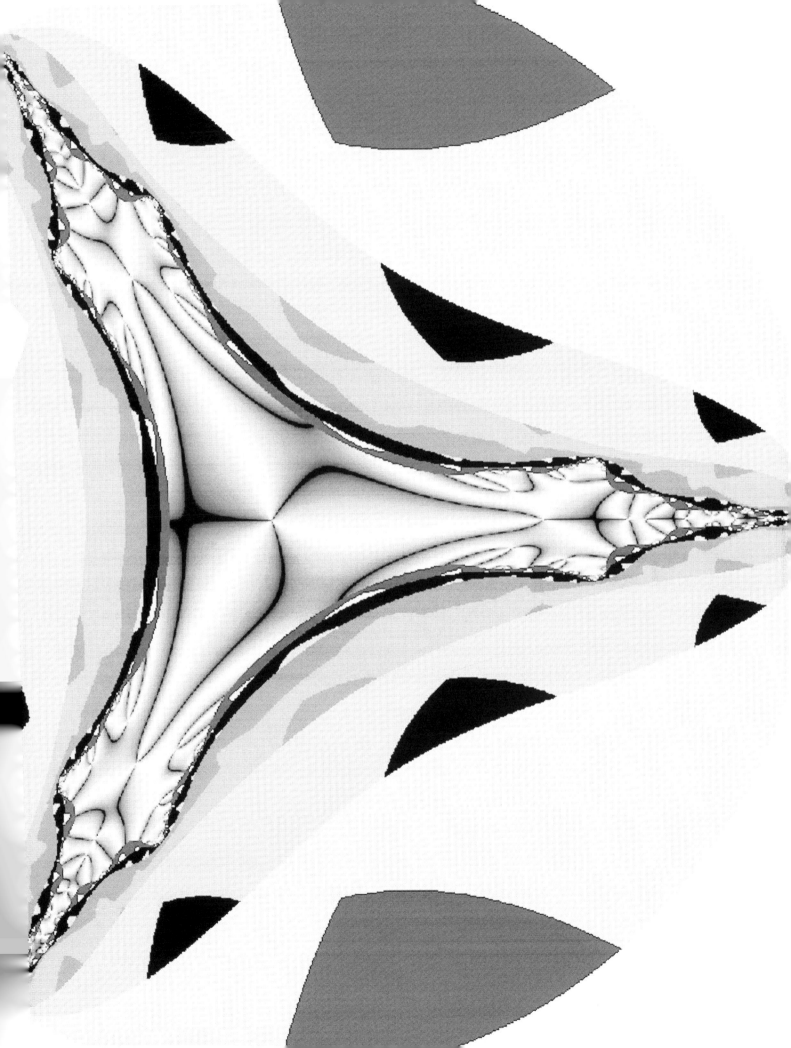

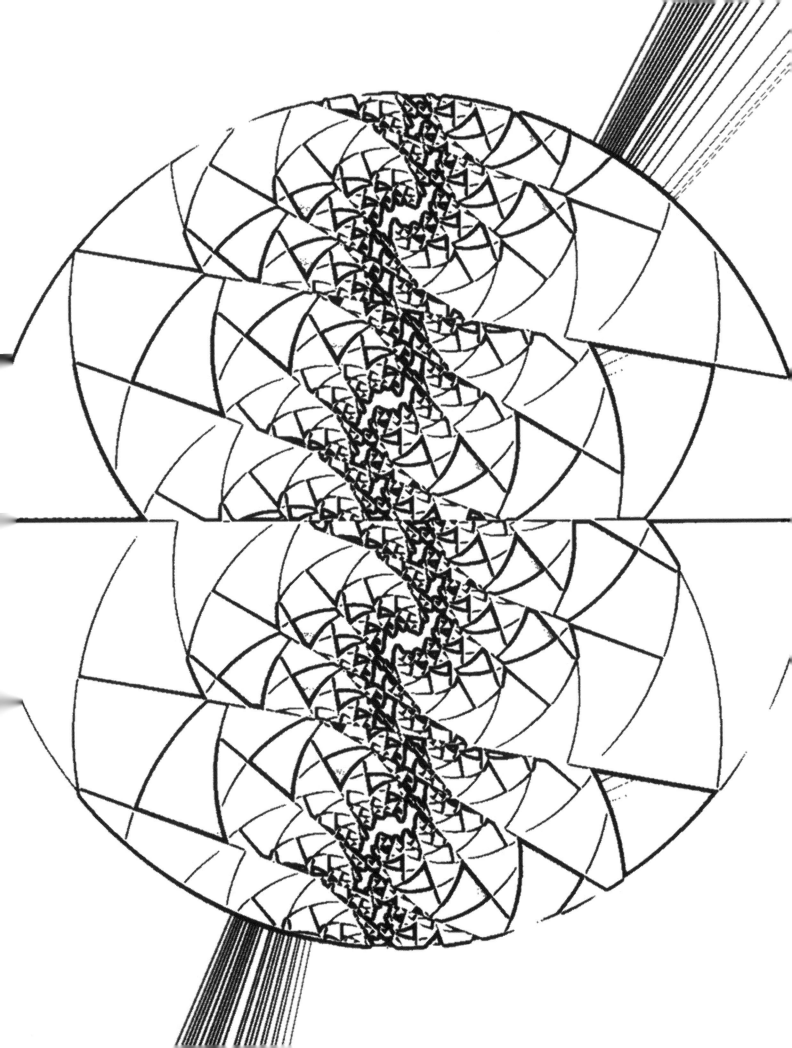

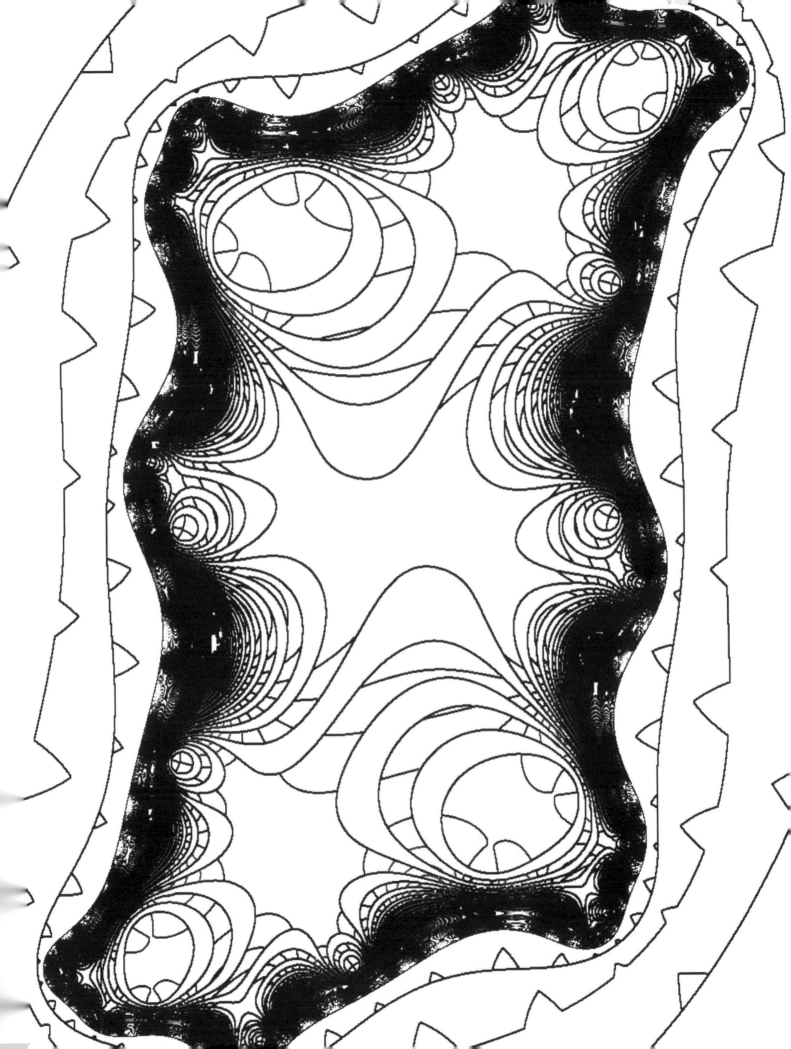

I hope you've enjoyed these designs!
To see more of my designs and products, please visit
www.FractalsBy5Leaf.com
www.Facebook.com/FractalsBy5Leaf

Made in the USA
Middletown, DE
26 November 2016